*An Interpretive Guide to the Exhibition Mounted at The N
to Accompany the XVth International Conference on t,
Cartography, June 1993*

TWO BY TWO

*Twenty-two Pairs of Maps from the Newberry Library
Illustrating 500 Years of Western Cartographic History*

Catalog prepared by James Akerman, Robert Karrow and David
Buisseret, and published with the help of Roger Baskes, Gerald
Fitzgerald, Arthur Holzheimer and the Chicago Map Society

The Newberry Library, Chicago, 1993

Table of Contents

Introduction

Instruments on Loan from the Adler Planetarium

List of Maps

Introduction

When the International Society for the History of Cartography accepted our offer to host its fifteenth conference in 1993, we knew that we were also accepting the challenge of mounting an exhibition of maps from the Newberry's collections. We thought at first of concentrating on maps of America, one of the conference themes, and a field in which the Newberry's collections are extraordinarily rich. But since so many of those attending the conference would be visiting the Library for the first time, and since our cartographic collections are in fact much wider in scope than merely the New World, we decided to attempt a broad overview of the collections. The task of selecting some fifty items to represent over five hundred years of cartographic history is daunting and perhaps foolhardy. But even more uncomfortable was the prospect of simply lining up our "treasures" and describing them, one by one, in chronological order; we feared that our specialized audience would see too much that was familiar and that our storyline would necessarily become the standard Whiggish paean to progress.

Our solution to this dilemma was to proceed two by two. Each map or atlas in the exhibition (and it includes a good many of our "treasures") has been paired with, and compared to, another map or atlas. We have tried to make our sample reflective of both the breadth and depth of our collections, but always we have made the maps speak in pairs. Some of these dialogues may seem more natural than others, but all, we trust, will illuminate aspects of each map that might not be obvious if either were studied in isolation. We hope that we have illustrated not only change over time, but also persistence over time, as cartographic topics, techniques, genres, styles, and ideologies are replicated in settings separated by hundreds of years and thousands of miles. We ask you to look at them not in the usual way, but "two by two."

In organizing this exhibition, we have incurred many obligations towards our colleagues at the Library. John Aubrey has given us much useful advice, and also detected several errors in the catalogue before they went public; Arthur Holzheimer, longtime Smith Center volunteer, also saved us from some mistakes. Kenneth Cain went to much trouble to ensure that all the photographs in this catalogue were as sharp as possible, and the conservation staff, headed by Joan ten Hoor, worked long and hard in conserving the artefacts and preparing them for exhibition.

Outside the Library, our chief thanks go to Mr. and Mrs. Webster, of the Adler Planetarium, who not only provided some artefacts for the exhibit, but also organized their own exhibit at the Planetarium for us, with the help of Liba Taub and Kate Desulis. We hope that they, and our visitors, will enjoy the exhibit as much as we enjoyed putting it together.

The illustration contains manuscript text in an old Italian hand, reading approximately:

S choglì fon molti perlomar coperti
fu tuperquote et rompe molte uolte
chi nonna i marinai ben defti et fperti
ifole grande et picchole fon molte
et deffe parleremo inluoghi cierti
quando uerremo ta oue fon uolte
ueggiamo tmprima ingienerare la terra
come fi rifide et come il mar la ferra.

VN T. dentro ad un .O. moftral difegno
come intre parti fu diuifo ilmondo
et lafuperiore ye maggior regno
che quafi piglia lameta del mondo
afya chiamata elghambo uicto et fegno
che parte ilterzo nome dalfecondo
affricha dicho da europa il mare
mediterano tra effe im mezo appare.

Q vefto tondo non ue mezo la fpera
ma molto meno et tucto laltro e mare
et non ne tucta quefta faccia intera
arida terra ma danauichare
fi truoua incierti parti gran riuiera
che ben laterza parte de bagnare
dacqua infalata che uiene dalgran cierchio
cha tucta laltra terra fa coperchio.

---

1 Gregorio or Leonardo Dati, World Maps from "La Spera" (Florence, c. 1425) [Edward E. Ayer Collection]

2 Johannes Honter, Map of Part of Asia from the *Rudimenta Cosmographica* (Kronstadt, 1542) [Edward E. Ayer Collection]

Here are two examples of verse cosmographies, illustrated with maps. The "Spera" was probably composed about 1422, scholars disagree about whether it should be attributed to Leonardo or Gregorio Dati. Leonardo (ca.1365-1425) was a Florentine Dominican who became the General of his order in 1414. He published a number of sermons and, possibly, commentaries on Aristotle.

His brother Gregorio (1363-1436), was a merchant and public official who wrote an important history of Florence. It is tempting to think that the brothers may have collaborated on "La Spera."

At all events, it was an extremely popular work, and over 150 manuscript copies have survived. Besides the several cosmographical diagrams and the "T-in-O" map, manuscripts of "La Spera" contain marginal maps showing parts of the Mediterranean coast with distances marked between some of the major ports. These maps are remarkably similar in different versions of the manuscript and appear to have been considered an integral part of the work as "published." Casting information in the form of verse was common strategy for encouraging memorization, and both the poetic nature of the work and the large number of surviving copies suggest that it may well have

2

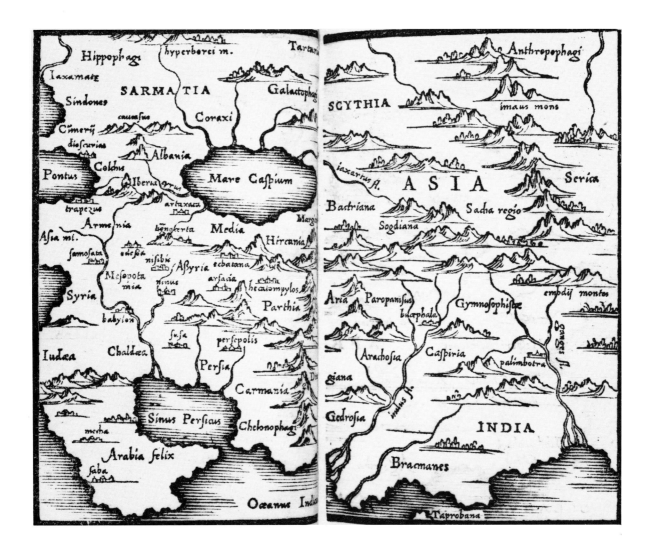

been used for didactic purposes.

There is no doubt about the educational intent of Johannes Honter's *Rudimenta Cosmographica*, first printed in Krakow in 1530. Honter (1498-1549) was born at the far edge of European culture, the German Transylvanian city of Kronstadt (now Brasov, Romania). He was probably educated by Dominicans in Kronstadt and went on to receive a master's degree from the University of Vienna. As a sort of adjunct professor at the University of Krakow, Honter published the first edition of the *Rudimenta*, in Latin prose, in 1530. The booklet was accompanied by two maps (the woodblocks cut by Honter himself), one of the world and one of the Eastern Hemisphere. Honter worked for a time in Basel as an editor, proofreader, and cutter of woodblocks.

He returned to Kronstadt in 1533 and immediately began to work toward fostering humanistic learning among the Transylvanian Germans. He brought in a printing press and types and produced the first printing in Kronstadt in 1535. Over the next few years he printed a number of small attractive editions of the classics for the use of students both at the Kronstadt Gymnasium and at a new school with a very advanced curriculum which he founded in 1545. In 1542 he printed the first edition (shown here) of a new version of his *Rudimenta* in Latin verse. It is a little longer than the "Spera" (1366 lines against 1152 lines in the Dati poem), but in organization and content there are strong similarities; it too is set in rhyme as an aid to memorization. At the end of the *Rudimenta* are 14 small woodcut maps, one of the world, three of Asia, one of North Africa, and nine of European countries and regions. These are modern and not Ptolemaic maps, and an argument could be made that they constitute the first modern world atlas.

References: Karrow and Vitti.

3

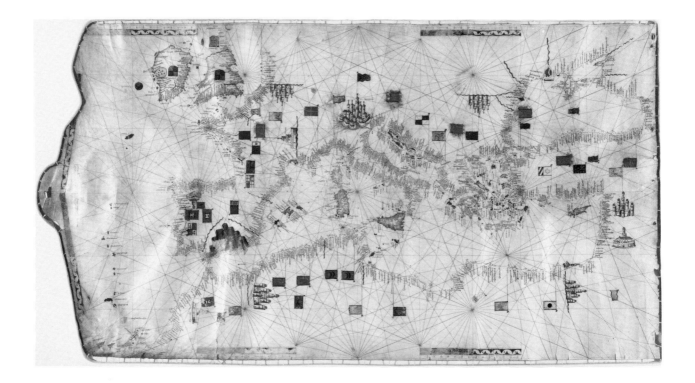

3   Petrus Roselli, Portolan Chart of the Mediterranean Sea (Majorca, 1456) [Edward E. Ayer Collection]

4   Edward Wright, World Map from Richard Hakluyt, *Principall Navigations, Voiages and Discoveries of the English Nation* (London, 1599) [Edward E. Ayer Collection]

Roselli's portolan chart is a typical product of fifteenth-century Majorca. Superbly drawn on vellum, it shows the outline of the coasts (but no interior detail) with ports named at right angles to them. Flags fly above the appropriate territories, and certain large cities are shown in miniature bird's-eye views. There is a spherical pattern of windroses, probably to help in the drawing of the chart, and four scales have been inserted. The elegance of this product suggests to many scholars that it was not meant, like some of its contemporaries, to be rolled on a spindle and carried on a ship at sea; no doubt it was intended for the information of some rich noble or merchant, to indicate the main trade centers in the Mediterranean Sea and a slightly wider world outside it.

At first glance, the world-map by Edward Wright comes out of another world. Vastly greater areas of the earth are shown, on the projection invented by Mercator that enabled such charts to be used as reliable navigating tools. The map is also *printed* on *paper*, which allowed its faithful duplication many times over, in sharp contrast to the vellum manuscript.

And yet, against these sharp differences, there are some curious continuities. Wright has kept the circular pattern of windroses, some of which have become rather elaborate compass-roses. He has retained the medieval way of delineating the coasts with exaggerated bays, and of placing the names at right angles to them. Like Roselli, he is not at all interested in interior details, except for the greatest rivers and lakes. Close inspection of both maps will show the use of stippling and small crosses, still used on nautical charts today to indicate shoal water and rocks.

In a deeper sense, too, these maps are cousins, for both are products of particular historical circumstances. Roselli wanted to show his patron what trading opportunities existed around the Mediterranean, and Wright wanted to demonstrate to his English readers that a whole world was opening up for the taking. They are thus also examples of the map as an instrument of power,

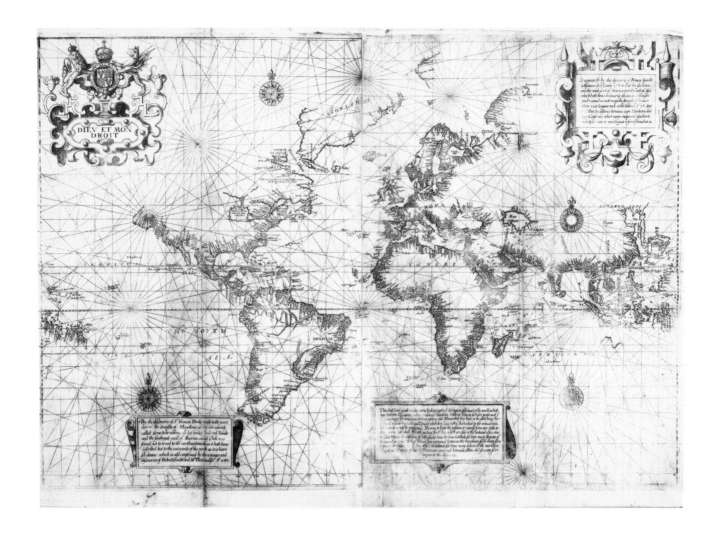

whether this power was projected around the Mediterranean Sea or around the whole world. The huge royal arms, at the top left on Wright's map, was thus emblematic of the dawning age of English worldwide sea empire.

References: Harley and Woodward, Karrow, La Roncière, Nordenskiöld and Shirley.

5 Erhard Reeuwich, "Civitas Venetiarum," from an early sixteenth-century edition of Bernhard von Breydenbach, *Peregrinatio in Terram Sanctam* [Franco Novacco Collection]

6 Jacopo de Barbari, Detail from *Venetie* (Venice, 1500) [Franco Novacco Collection]

These two views show Venice at the height of her power, towards the end of the fifteenth century. The long woodcut by Erhard Reeuwich comes from Breydenbach's *Peregrinatio*, and is the first of five such woodcuts in that book; they marvelously demonstrate the skill of the German craftsmen of this period. The city is seen almost at ground level, in what was called a "profile," and this allows Reeuwich to offer us a very realistic approach view, even if the Alps have been brought improbably close.

Saint Mark's cathedral, with its great tower, and the Doge's Palace ("palacium dogis") stand out, as do the many churches in the various quarters. Reeuwich had a sharp eye for things nautical, and the ships are observed with great fidelity. Note the huge galley at the jetty off the doge's palace, and the numerous "round ships" as they were called, the great bulk carriers of late medieval Europe. Two complete ones are to be seen at the extreme right of the view, and to the left of them a huge new ship is being built. Many other such ships under construction may be seen elsewhere in the woodcut, emblematic of the role of Venice as a great maritime power.

This role is even more explicitly stated in the detail from Barbari's great woodcut. Here Neptune reigns just off Saint Mark's cathedral and square. Barbari's extraordinary work, in six great sheets, shows Venice from a high angle, so that virtually all the houses and streets can be included. Even if his work turns out to be less precise than once was thought, it is a technical *tour de force*, well illustrating the superiority of the bird's-eye view over the earlier profile.

Obviously, Barbari is able to show more detail. But he is also able to bring out salient features which escape the profile altogether; for instance, he clearly shows the extent of Saint Mark's square, of which Reeuwich gives only a hint. The bird's-eye view would eventually be superseded for most purposes by the vertical planimetric image. But it remains popular to the present day for certain uses, including maps drawn for tourists, who by using it can almost instantly seize not only the general outline of a town, but also the location of its main monuments.

References: Breydenbach and Schulz.

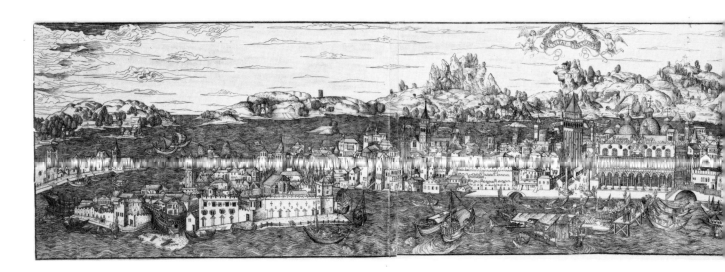

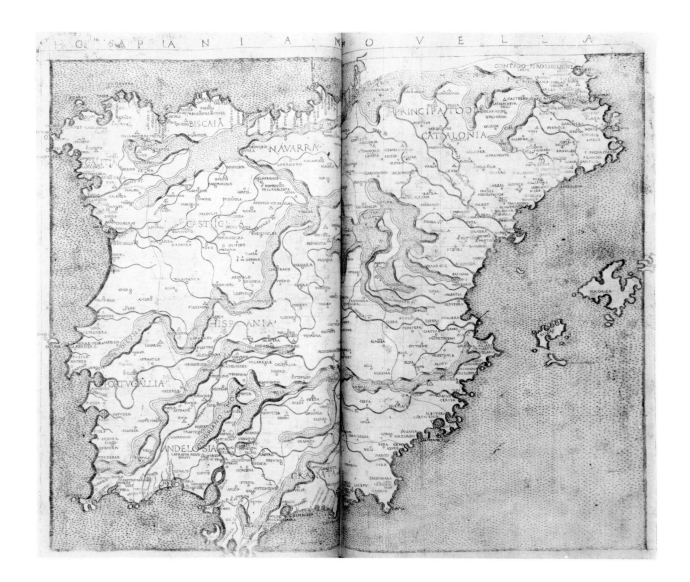

7 Francesco Berlinghieri, "Spania Novella" from his *Geographia* (Florence, 1482) [Edward E. Ayer Collection]

8 Joan Blaeu, "Tractuum Borussiae, circa Gedanum et Elbingam Delineatio," from the *Atlas Maior* (Amsterdam, 1662-67) [John M. Wing Collection

Each in their own time, these two atlases were the most comprehensive cartographic treatment of the entire world available in print to Europeans. Berlinghieri's *Geographia* is among the earliest of all printed cartographic works, appearing only ten years after the first European map to be printed (a "T-in-O" world map, published in a 1472 edition of the *Etymologiae* of Isidore of Seville). It is one of six editions of the maps from the *Geographia* of Claudius Ptolemy published before 1500, five of which the Newberry possesses as part of the Stevens-Ayer Ptolemy collection. Berlinghieri's text is, however, a contemporary geographical survey based partially on ancient sources. Alongside the usual 27 Ptolemaic maps are four *tabulae modernae* (Spain, France, Italy, and Palestine) based on those by a previous Ptolemaic editor, Nicolaus Germanus. We show here the map of "modern" Spain. Copper engraving was still in its infancy, and Berlinghieri's engraver seems not to have known how to correct the mistake in the title.

Sixteenth-century editors, such as Martin Waldseemüller and Sebastian Münster, added dozens of new maps, mostly of European regions, to their versions of Ptolemy. However, by the last third of the sixteenth century the circulation of printed maps of European, Asian, and American territories had expanded tremendously, so that Ptolemy's framework was no longer suitable for a contemporary world atlas. The introduction of entirely new world atlases like Abraham Ortelius's *Theatrum Orbis Terrarum* or Cornelis de Jode's *Speculum Orbis Terrarum* (item 19) nevertheless did not halt this cartographic inflation. In the seventeenth century, new descriptions of China, the Indies, America, and sub-Saharan Africa streamed back to Western Europe to be snatched up by the now very competitive atlas industry.

The rivalry of the Amsterdam houses of Blaeu and Hondius-Jansson was particularly fierce; each freely copied the maps of the other while adding entire volumes of new maps and textual descriptions. By the mid-1650s each house had put out similar six-volume editions of their *Atlas Novus* or *Novus Atlas*, published in several languages and containing more than 550 maps. We display a magnificent colored map of East Prussia from the ultimate Blaeu atlas, *Atlas Maior*, (12 volumes, published 1662-67). Blaeu's atlas not only encompasses a wider European world than the Berlinghieri atlas of nearly two centuries earlier, but illustrates two centuries of refinement in the art of printing maps from copper plates. This copy of the *Atlas Maior* is bound in its original red velvet, reserved for important customers or those able to pay the highest premium. Alas, the Newberry's Wing Collection possesses just two volumes of this exquisite copy, but a complete copy of the 1663 French edition may be found in the Ayer Collection.

References : Koeman (1970), Koeman and Homan

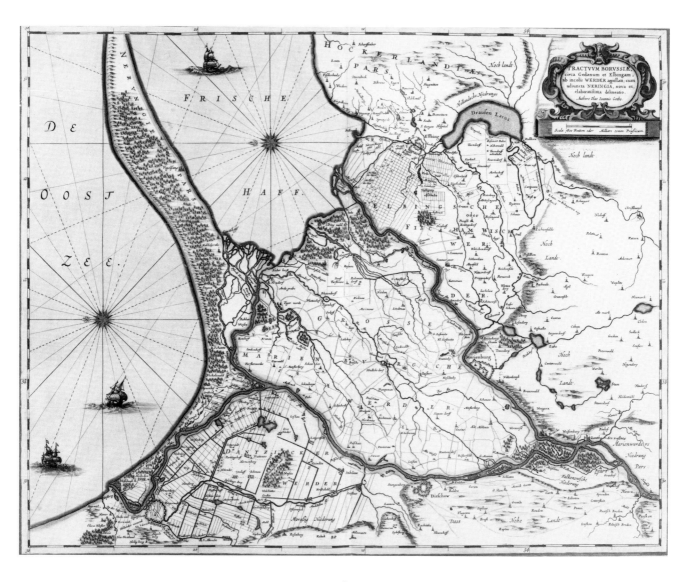

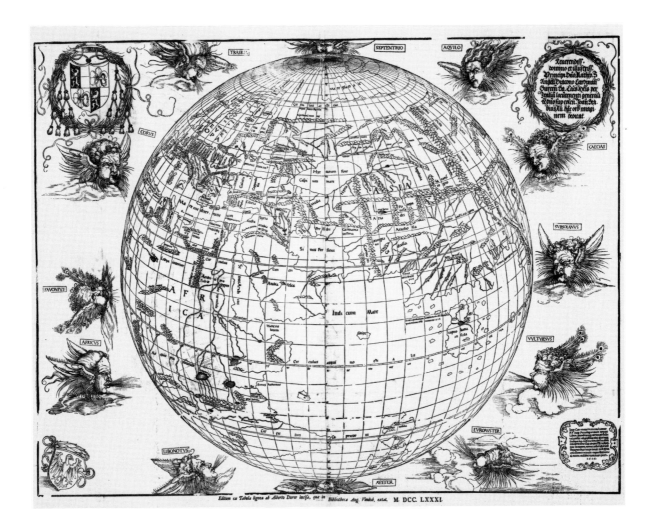

9 Albrecht Dürer/Johannes Stabius, Hemisphere (Nuremberg, 1515) [Franco Novacco Collection]

10 Oronce Fine/Hadjii Ahmed, World Map (Venice, 1566) [Franco Novacco Collection]

This pair of maps exemplifies the sixteenth-century fascination with map projections and the work of the Austrian mathematician and geographer Johannes Stabius. Medieval artists had had no satisfactory method of distinguishing between a disc and a sphere; looking at a T-in-O type map, for instance, we have to make an effort to remember that its author intended it to represent the half of the globe taken up by land, and that the other hemisphere ("behind" the one shown) was wholly water.

When we look at Johannes Stabius's map, we can see at once that it represents one half of a globe, for that is exactly what it looks like. The map, published in 1515, is the first known use of the orthographic projection for a world map. It recreates, in effect, the appearance of a globe and there is a good likelihood that Stabius based his map on an actual globe, Martin Behaim's, made in Nuremberg in 1492. Stabius's map was probably meant more to celebrate the possibilities of the mathematical transformation of space, as a *trompe l'oeil* novelty, than as a working geographical tool. By providing a sense of what the earth would look like from space, it was the sixteenth-century equivalent of the "whole earth" images that emerged from the American space program in the 1960s.

Although the woodcutter's name is not given, the map can be confidently attributed to Albrecht Dürer who, besides being an extraordinarily talented artist, was fascinated by mathematics and geometry and wrote a treatise on how to represent various kinds of solid shapes in perspective.

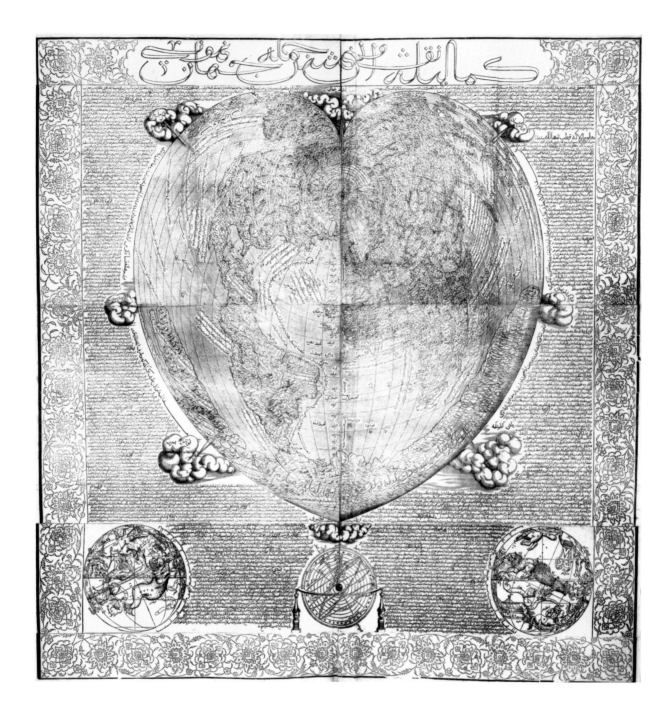

Understanding perspective, he wrote, was "like seeing a new kingdom."

The heart-shaped, or cordiform projection, was also invented by Stabius. Although novel in appearance, the cordiform is, in fact, an equal-area projection. The path from Stabius in Vienna to this Turkish map is a bit circuitous, but its immediate source seems to have been a map drawn by the French mathematician and geographer, Oronce Fine, and presented to his king in 1519. The "cosmographic heart" may have been an attractive emblem at the French court where the Sacred Heart devotion was becoming significant.

This Turkish version was commissioned by Bajazet, a son of Suleiman the Magnificent, in 1554. Neither of these maps exists in sixteenth-century impressions; both sets of woodblocks were discovered in the late eighteenth century and all existing impressions were made thereafter.

References: Brown, Karrow and Kish.

11 Ottomano Freducci, The Central Mediterranean Sea from a Portolan Atlas (Ancona, c. 1533) [Edward E. Ayer Collection]

12 Sebastião Lopes, The Indian Ocean from a Portolan Atlas (Lisbon, c. 1565) [Edward E. Ayer Collection]

These atlases emerge from the tradition of portolan chart-making that was already at least two centuries old at the dawn of the sixteenth century. Each is drawn in several colors on vellum, and has the characteristic system of rhumblines and names of major ports inked in black and red perpendicularly to the scalloped coastlines. The similarities end there.

Despite its date of manufacture, the Freducci atlas is essentially a compendium of medieval knowledge of the Mediterranean Sea. One could without difficulty disassemble the atlas and remount its ten panels as a single "normal portolan" chart, of which our exhibit 3 is an example. Its still bright and sharp lines tell us that it never went to sea, but resided in someone's library - perhaps that of an Italian patrician. Nine surviving atlases very much like this one (dated 1497-1539) attributed to Freducci survive, suggesting that his family traded profitably in these otherwise undistinguished atlases.

The magnificent atlas of 24 charts by the Lisbon cartographer Sebastião Lopes has wider horizons. It takes its reader on a delectable tour of the entire world as then known to maritime Europeans. The itinerary, commencing with Atlantic Europe, Brazil, and the circumnavigation of Africa, recapitulates Portuguese maritime history. The volume is opened midway on its traverse of the Indian Ocean to charts of the Arabian Sea. Miniature views of three important entrepôts, Jidda, Ormuz, and Goa underscore their economic importance to the Portuguese. This, and the prominence of the Portuguese arms, suggest that we have here a gilded reference atlas made for someone quite close to the center of Portuguese power.

References: Cortesao and Teixeira da Mota, Errera, Harley and Woodward, and Smith.

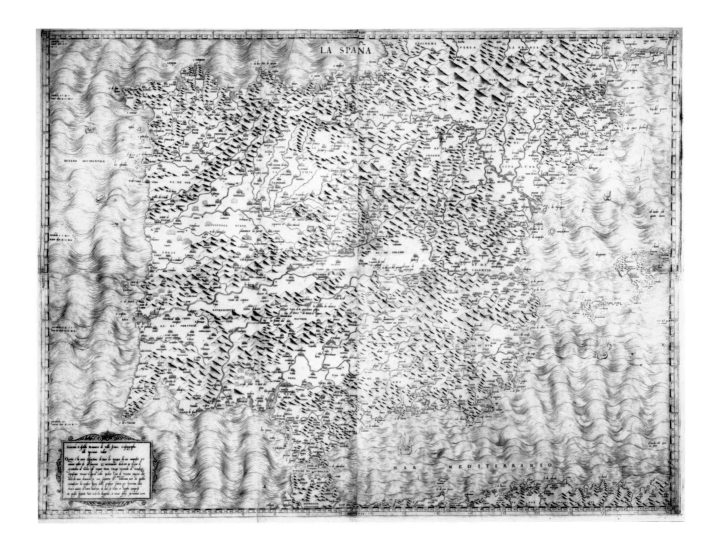

13 Jacopo Gastaldi, *La Spaña* (Venice, 1544) [Franco Novacco Collection]

14 Philip Apian, Index Sheet from the *Bairische Landtafeln* (Ingolstadt, 1586) [Edward E. Ayer Collection]

The two maps shown here emerge from a period of intense development in the art of cartography in Europe, and they interestingly illustrate old and modern elements. Gastaldi's map, for instance, is a superb "modern" copper-engraving, whereas Apian's key-map to Bavaria was engraved in wood by Jost Amman in the old German tradition; the 24 blocks needed for this are still preserved at the National Museum in Munich.

However, if Apian's map is rather old-fashioned in its mode of production, it is highly innovative in its cartographic content. This is one of the earliest examples of a key-map, in which the ensuing contents of a uniform atlas are set out at the start, using numbered blocks. The key-map and "cellular" multi-sheet map have become quite familiar to us, but it required a mastery of different scales, and a general understanding of the whole extent of the territory involved, that were hard to attain. Gastaldi's map, on the other hand, shows Spain on four sheets, and is one of the great achievements of the high period of Italian copper-engraving, when disparate maps were produced to make up what the French call *atlas factices*, or atlases composed of maps of a variety of sizes. It is Gastaldi's earliest known production.

Two years after Apian's Bavarian atlas came out, Ortelius published his *Theatrum Orbis Terrarum* (Antwerp 1570), in which maps of the whole world were bound together at a uniform size, having been specially commissioned for this purpose. The *atlas*

14

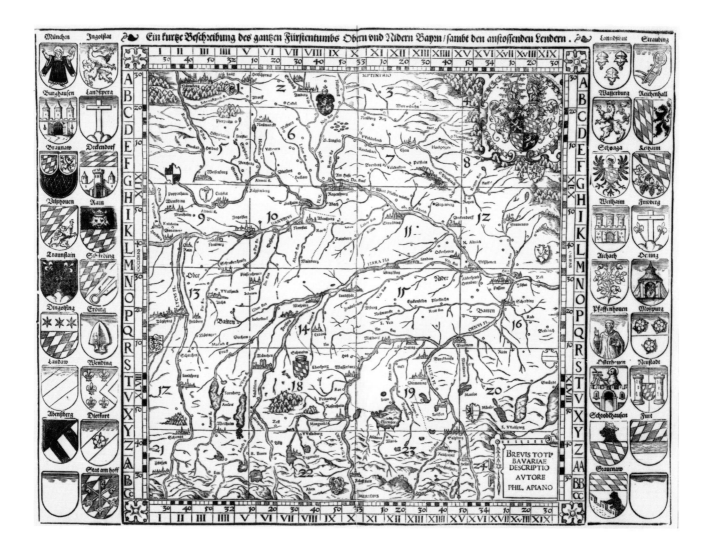

factice thus became outmoded, and atlas-production entered upon a new era, using the printing-technique of Gastaldi and the atlas-arrangement of Apian.

References: Almagiá, Karrow and Wolff.

S. DOMINGO.

15 Baptista Boazio, "Civitas S. Dominici sita in Hispaniola," from Walter Bigges, *Expeditio Francisci Draki* (Leiden, 1588) [Edward E. Ayer Collection]

16 François Olivier, "Carte du Fort et Bourg de Paramaribo," 1724, from the "Cartes Marines" (c. 1715-1725) [Edward E. Ayer Collection]

The age of European overseas expansion was marked by intensive cartographic activity, as rulers tried to bring "their" territories under intellectual control. They were also very interested in obtaining knowledge about the overseas possessions of other European powers, and here we have an English map of a Spanish city, followed by a French map of a Dutch settlement.

The map by Baptista Boazio, an Italian mapmaker living in London, is one of four accompanying an account of Sir Francis Drake's piratical voyage to the New World in 1585-86. Drake and his men captured Santo Domingo on New Year's Day, 1586, and extorted a huge ransom. Boazio's map (possibly copying a map by Drake himself) shows the city, with Drake's fleet lying offshore. Boazio sets out the area in a bird's-eye view, that allows him both to show the characteristic street-pattern, with the main

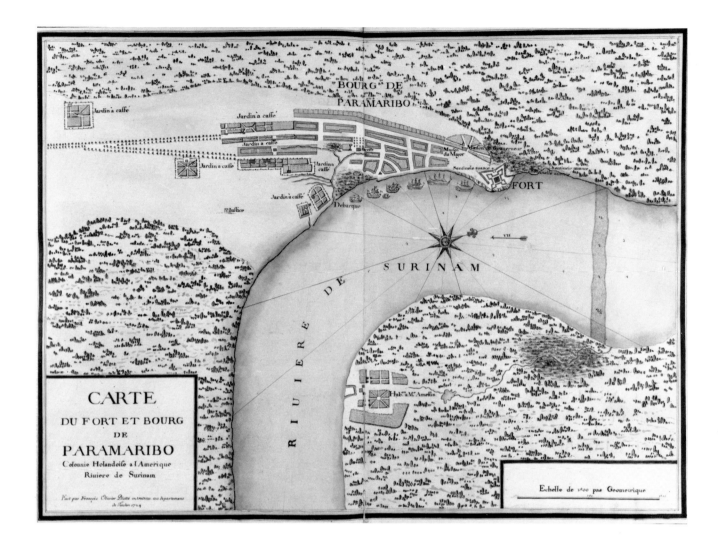

buildings clearly indicated, and to give some idea of the surrounding topography. There is a mark of orientation, but no scale; this plan is essentially an informal layperson's view. It does not shy away from fabulous fauna and its legend celebrates the successful English siege.

The Dutch colony of Surinam in the late seventeenth century enjoyed considerable prosperity. Huguenot and Jewish refugees (some driven from the neighboring French colony of Cayenne) helped to establish the coffee industry, which became a mainstay of the economy. In 1712, a French fleet from Cayenne under the privateer Jacques Cassard managed to subdue the colony and extract an enormous levy from it.

The plan of Paramaribo by François Olivier, "Pilot of Toulon," is dated 1714 but was most likely made during Cassard's three-month occupation in 1712. It exemplifies the next generation of colonial maps. Only the ships are shown in bird's-eye view now, and exotic creatures are left to the imagination. The rest is delineated according to the scale, which would allow us precisely to estimate the size of the coffee-plantations ("jardins a caffé").

This map is one of the 115 "Cartes Marines," a uniform collection of maps, made about 1725, and drawing on French official sources. It seems to have been assembled in Switzerland, perhaps by some Huguenot residing there, and contains much unique material.

References : Buisseret (1991), Cohen and Kraus.

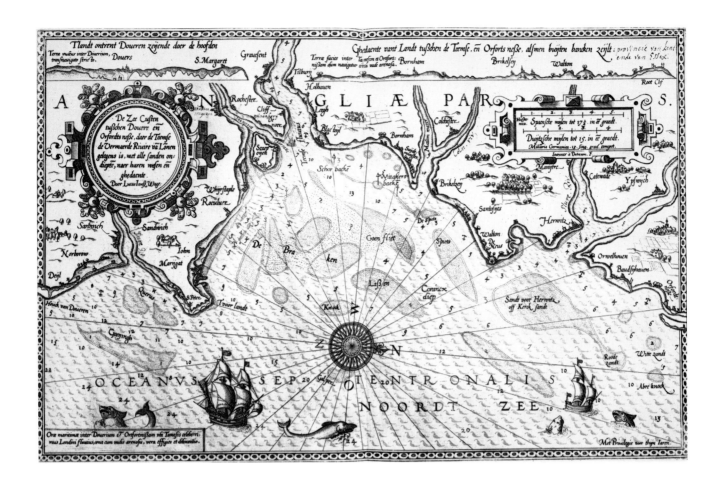

17 Lucas Jansz Waghenaer, "De Zee Custen tusschen Douvre en Orfordts Nesse," from the *Spieghel der Zeevaerdt* (Leiden, 1588) [Edward E. Ayer Collection]

18 Approaches to the Thames Estuary from *The English Pilot*, Part II (Southern Navigation)(Dublin, 1766) [Edward E. Ayer Collection]

The two atlases shown here represent different stages in the cartographic delineation of European waters. The first edition of the *Spieghel* (in English, *Mariner's Mirror*) came out at Leiden in 1584-5, and was the equivalent for marine charts of the 1570 Ortelius atlas for general maps: that is to say that for the first time it reproduced a considerable body of charts in a uniform printed atlas. The *Spieghel* went through many editions in Dutch, French, Latin and English, the last appearing in 1622. About fifty years later, John Seller published the first volume of

the *English Pilot*, another marine atlas which also had a long life, appearing in various guises and with various publishers far into the eighteenth century.

We here show Part II of the *English Pilot*, dealing with navigation around England and Southern Europe. The map, which is oriented westward, as if for a mariner entering the Thames estuary, shows the east coast of England from Orford to Dover, and the detail on plate 18 shows the right-hand side of the map. Plate 17 shows the same area from the corresponding map in the *Spieghel*; note "Orford" at the extreme right of both plates and "Colchester" at the top left.

The *English Pilot* version of this complicated stretch of sea is much more detailed than had been *Spieghel's*; sand banks are now shown much more precisely and there are many more soundings. All this reflects the greatly increased activity off this part of the English coast during the seventeenth century, when London was becoming a great port both for local and for world trade. Both maps show clearly their roots in the portolan chart tradition.

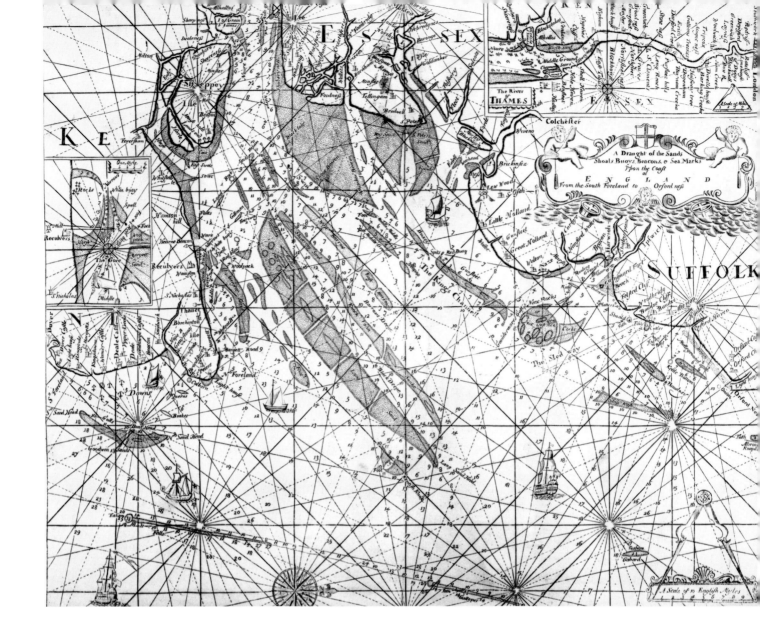

The compass roses and radiating rhumb lines provide the sailor with his bearings through trackless waters. Except for coastal details, and important towns, the land surface is *terra incognita*. Shoals and rocks are shown with the time-honored symbols.

A relatively recent innovation is the coastal profile, to help the sailor recognize a coast. In Waghenaer, the general profiles about the map attempt to show the general sweep of the land, while on the map proper, the coasts themselves are running profiles. Like portolans, too, the orientation of the chart is a matter of convenience, not convention. Typically the chart is arranged on the page so as to provide a view looking landward, and the north arrow falls where it will.

Hydrographic surveying had made great progress in the almost 200 years separating these charts. The *English Pilot* exhibits many more soundings and a more detailed picture of the sand banks. The first detailed manual of hydrographic surveying, Murdoch Mackenzie's *Orcades*, dates from about this time.

References: Howse, Koeman and Homan (1967/1985), Skelton (1964) and Waghenaer.

19 Cornelis de Jode, "Croatiae & Circumiacentiu Regionu versus Turcam Nova Delineatio," from Gerard and Cornelis de Jode, *Speculum Orbis Terrarum* (Antwerp, 1593) [Edward E. Ayer Collection]

20 Johann Baptist Homann, "Franckfurt am Mayn mit ihrem Gebiet," from an *Atlas Factice* (c. 1720-1750) [Sack Collection]

These items represent two traditional ways of putting maps together in bound form. Gerard and Cornelis de Jode's *Speculum Orbis Terrarum* is the older of the two atlases, but is, in many ways, more like our modern atlases. It has an elaborate printed title page and introductory text, followed by 85 maps engraved in a uniform style, each introduced by a brief textual description of the various countries and regions mapped. Though it lacks a proper table of contents, it has an extensive geographical index at the back, which tells us that the editors intended each copy of the atlas to be identical to every other in content and arrangement. Large standardized atlases of copper-engraved maps like this one were expensive to make, which may partially explain why the *Speculum* saw only two editions (1578 and 1593) before quietly yielding to the better reputed atlases of Gerard Mercator and Abraham Ortelius.

Our second atlas is an oversized volume that once belonged to baron Johan Gabriel Sack (1697-1751), a Swedish nobleman and armchair geographer. Sack had family ties to Swedish diplomatic circles, and his map collection suggests a learned man avidly interested in European affairs. The maps were probably collected during the decades of Sack's adult life (c. 1720-1750), and include work by all the major publishers in France, Germany, and the Netherlands, as well as many minor publishers. The heart of the collection is the roughly 600 atlas-size maps that he had bound in two large volumes. Though the other of these volumes has now been disbound, this one has been preserved intact to demonstrate how a wealthy eighteenth-century collector kept his maps. The unusually large format of the atlas allowed each map to remain unfolded; this accounts for the nearly pristine condition of the brilliantly colored maps.

This Sack volume, then, is among the finest surviving examples of a composite atlas, a book of maps selected and arranged for the convenience of an individual collector. This way of manufacturing an atlas was still common in the early eighteenth century, and seems to have been encouraged by many map publishers, since it spared them the expense of incorrectly guessing public demand for atlases of a specified size and content. The practice died away only in the nineteenth century, when the introduction of lithography to map printing made it possible for publishers to mass-produce inexpensive standardized atlases tailored to various tastes and pocketbooks.

Both atlases are especially well endowed with maps of the many provinces and petty sovereignties of Central Europe, and we offer for comparison, J. B. Homann's map of the city-state of Frankfurt-am-Main and de Jode's rendering of Croatia.

References: Bosse, Koeman and Homan, and Skelton (1965).

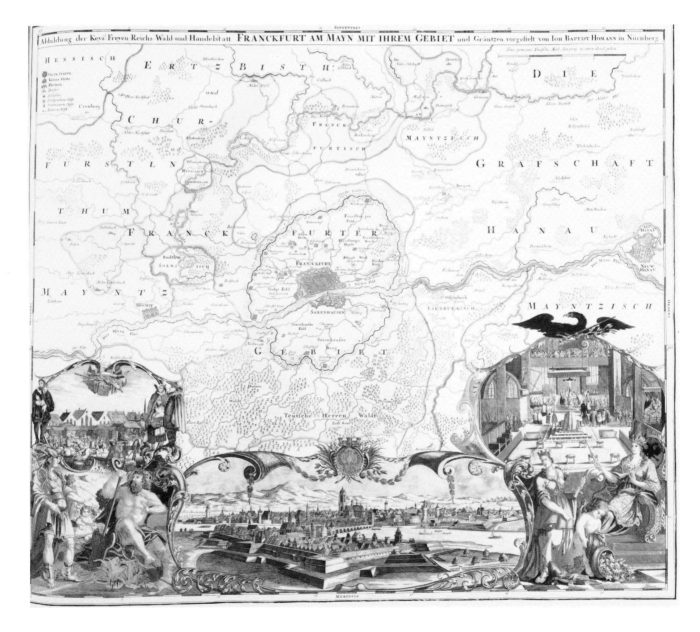

21

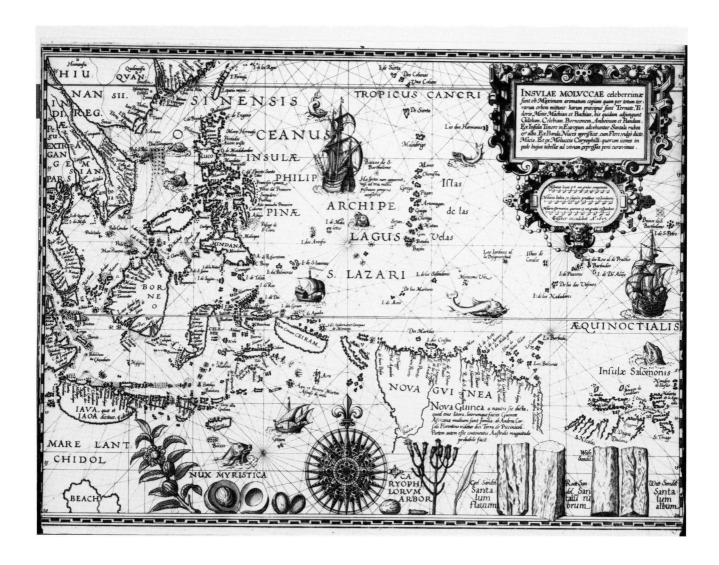

21 Bartolomeo Lasso, *Insulae Moluccae* (Amsterdam, 1592-94) [Visscher Collection]

22 Samuel de Champlain, *Carte Géographique de la Nouvelle France* (Paris, 1612) [Edward E. Ayer Collection]

T he "great age of exploration" witnessed the opening of large parts of the world to western eyes. Not only did Europeans learn, from the explorers' maps, about new coastlines and strange place names; they also acquired, through the accounts of travelers, information about new and exotic people, animals, and plants. Here are two explorers' maps that pay special attention to the unusual botanical riches associated with the new lands shown.

Petrus Plancius, whose map of the Moluccas is

on the left, was the official cartographer to the Dutch East India Company from 1602 to 1619. He somehow got access to manuscript maps by the Portuguese cartographer Bartholomeu Lasso, and printed a number of them in 1592. Lasso worked in Lisbon producing nautical charts and would have been privy to the latest information from Portuguese captains and pilots. The Portuguese had reached India in 1498, and the Moluccas (otherwise known as the Spice Islands) in 1511. By the time of Plancius's map, the Portuguese had enjoyed almost a century of virtual monopoly on the trade from the far east, a monopoly that the Dutch were ready to challenge; by 1599 there was a Dutch settlement in Molucca.

Three of the most important exports from the far east are illustrated on Plancius's map: *Nux myristica* (nutmeg), *Caryophilorum arbor* (cloves) and three varieties of *Santalum* (sandalwood). Plancius

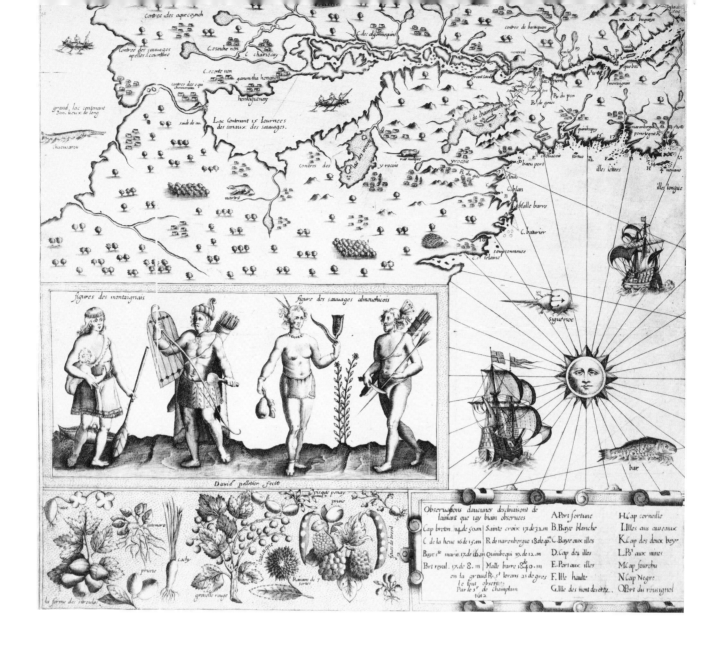

would have taken his illustrations from any of the numerous illustrated herbals that had been published during the course of the sixteenth century. In the case of his drawing of the clove plant, we can say for certain that his source was a woodcut in Charles de l'Escluse's edition of a book on the aromatic and medicinal plants of the Indies written by the Portuguese Garcia da Orta.

Samuel de Champlain, the "Father of New France," made the map on the right to illustrate his *Voyages*, published in 1613. Besides providing the first detailed picture of the topography of the St. Lawrence Valley and the first cartographic evidence of knowledge of Lake Superior (the "Grand Lac" at the left edge of the sheet), Champlain's map includes a wealth of ethnographic and botanical information. Among the plants he depicts we can identify squash (*sitroules*), plum (*prune*), red currants (*groiselle rouge*), chestnut (*chataigne*), three kinds of grapes (*raisains de 3 sortes*), and red beans (*foues de bresil*). One of the Indian women holds a squash and an ear of corn, and below her feet is an illustration of the Indian potato (*pisque penay*). Potatoes, squash, and corn were unknown in the Old World before the Columbian encounter, but thanks to systematic botanists like l'Escluse, they were soon grown in special botanical gardens and eventually found their way into the European diet.

These explorers' broadsides, combining cartographic with other kinds of information, foreshadow the kind of encyclopedic treatment of flora, fauna, and ethnography that characterized the work of the great eighteenth- and nineteenth-century voyages of exploration.

References: Ganong, Heidenreich and Pickering.

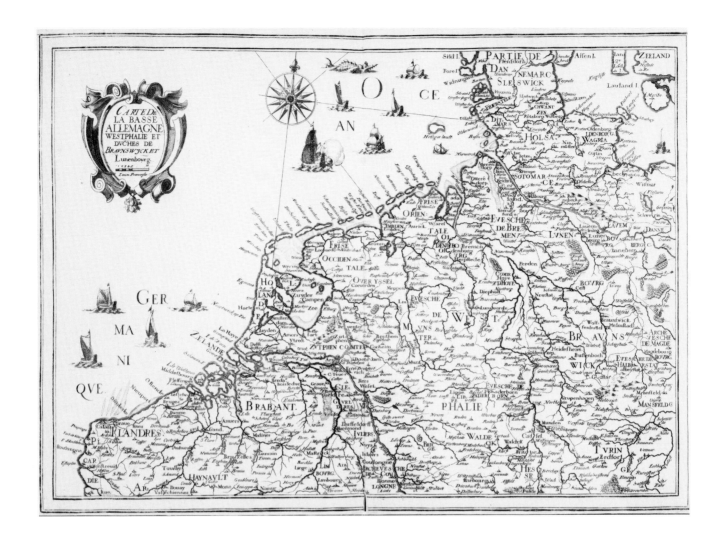

23   Christophe Tassin, "Carte de la Basse Allemagne," from *Les Cartes Générales de Toutes les Provinces de France, Royaumes et Provinces de l'Europe* **(Paris, 1637-[45?]) [General Collections, purchased from the Andrew McNally Fund]**

24   Nicolas Sanson, "L'Afrique, ou Lybie Ulterieure," from the *Cartes Générales de Toutes les Parties du Monde* **(Paris, 1658-[59]) [General Collections, purchased from the Andrew McNally Fund]**

From the time of Ortelius to the middle of the sixteenth century, atlas publishing in Europe was dominated by the Flemish and the Dutch. However, beginning in the 1630s, French publishers made a number of innovations in atlas design and structure that assured their leadership by the end of the century.

We show here two of the more innovative French atlases from mid-century. Christophe Tassin was a royal engineer who also published a number of modestly successful atlases of French provinces, cities, and coastlines during the 1630s and 1640s. The *Cartes Générales* (first published in 1634) was his grandest atlas project, but had the misfortune of appearing shortly after Melchior Tavernier's *Théâtre Géographique du Royaume de France* (1632, etc.), made up of maps first published by the Amsterdam house of Hondius-Jansson. Tassin's maps were closer to recent manuscript sources than Tavernier's. Nevertheless, the *Cartes Générales* was less attractive than the *Théâtre* and a commercial failure that little influenced subsequent published cartography outside of France.

The *Cartes Générales* is notable for its incorporation of nine large maps on multiple sheets, chiefly of countries or regions beyond France's borders. Here, for example, is the

northwestern sheet of Tassin's map of Germany in four sheets. Though the outlines of the modern Low Countries, northwestern Germany, and southern Denmark are clearly recognizable, this sheet does not focus, as most previous atlas maps did, on one or two particular provinces or countries. This atypical structure - before Tassin, only Gerard Mercator had employed this method so extensively - had the advantage of preserving the mapping scale throughout large portions of the atlas.

Nicolas Sanson's is the most important of all seventeenth-century French atlases and is arguably the most innovative atlas of its century. Sanson was a *géographe de cabinet*, an office-bound geographer who made his living digesting, appraising, and reorganizing geographical information from field and published sources into more useful forms. Sanson's maps are sometimes criticized as being unoriginal and regressive in terms of their geographical content, but this misses the aim of his work entirely. While most previous atlas editors had been content to draw their maps rather uncritically after whatever images happened to be available to them, Sanson thoroughly reworked the geography of each and every part of the world he mapped. Then each map was fitted into a hierarchical schema of the world's divisions, which he first laid out in a set of synoptical tables published in 1644.

Colored lines of varying hues and thickness are carefully coordinated with variations in the size and style of regional place-name labels. The largest regional divisions (e.g., "Pays des Negres") are named in the largest lettering and bounded by the thickest lines, while smaller lines and lettering are used for regional subdivisions (e.g., "Agades Regnum"). We take for granted in our atlases this standardization of regional values and symbols, which imparts apparent order to the often disorderly political world, but Nicolas Sanson was the first to apply such a system throughout a large atlas. It is no coincidence that this innovation came at a time when the legal and philosophical basis of the modern system of sovereign territorial states was being worked out.

References: Koeman (1963) and Pastoureau (1984 and 1988).

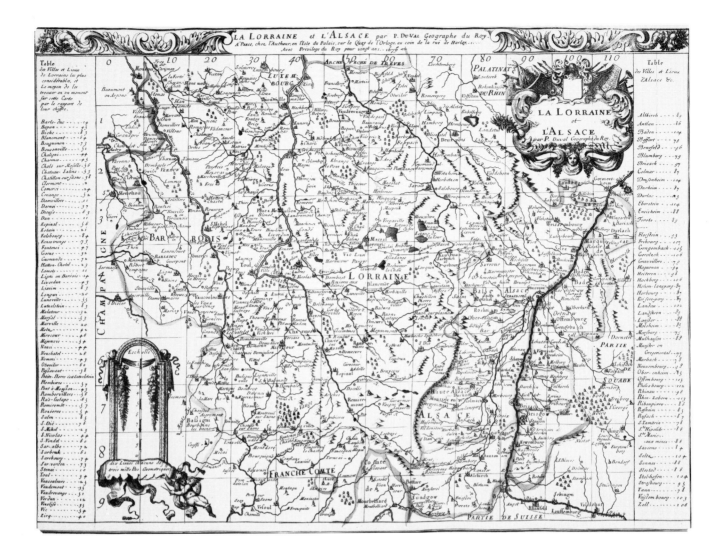

La Lorraine et l'Alsace par P. Du-Val Geographe du Roy.
A Paris, chez l'Autheur, en l'Isle du Palais, sur le Quay de l'Orloge, au coin de la rue de Harlay.
Avec Privilege du Roy pour vingt ans. 1676

LA LORRAINE et L'ALSACE par P. Duval Geographe du Roy

25 Pierre Duval, "La Lorraine et l'Alsace," from the *Cartes de Géographie les Plus Nouvelles et les Plus Fideles* (Paris, 1679-[86]) [Edward E. Ayer Collection]

26 Sébastien de Pontault de Beaulieu, "Pont a Monson," from the *Plans et Cartes des Villes d'Artois* and *Les Plans et Profils des Principales Villes des Duchez de Lorraine et Bar* (Paris, n.d.) [General Collections, purchased from the Andrew McNally Fund]

These two atlases emerge from the highly innovative period of French cartography that developed during the reign of Louis XIV (1661-1715). The Sun King was not the first sovereign to appreciate the administrative, military, and political value of maps, but no European monarch before him had organized and supported the cartographic activities of the state so extensively. During the 1660s the first of the great national topographic surveys based on triangulation was conceived and inaugurated by Jean-Dominique Cassini and Jean Picard with the support of the powerful minister Colbert. In the king's name Colbert also enlarged the system of royal engineers, made provincial intendants responsible for the acquisition and organization of the existing maps of their territories, and put the first *géographe du roi*, Nicolas Sanson, in charge of compiling this information into useful reference maps.

One significant by-product of extensive state cartographic sponsorship was the flowering of the Parisian map trade under the leadership of Sanson, Pierre Duval (Sanson's successor as *géographe du roi*), and men such as Sébastian de Pontault de Beaulieu, who had been a military engineer under Louis XIII. Commercial cartographers paid their

debt to the state with the publication of beautiful atlases that documented and publicized the king's many conquests along France's frontiers. The best-known of these was the series of atlases of Beaulieu's plans and views of towns in frontier regions. Two of these "petits Beaulieus," Artois and Lorraine, are bound together in the volume shown, which is opened to Beaulieu's view of Pont-à-Mousson, on the Moselle in northern Lorraine. The "petits Beaulieus" were extremely popular and served as a seventeenth-century form of propaganda legitimizing and celebrating French conquests.

Like Beaulieu, Pierre Duval published several popular atlases showing theaters of war, including maps of newly conquered territories and fortified cities. His much larger *Cartes de Géographie*, one of the first French world atlases, is a more subtle form of propaganda. Though its scope is global, its perspective is decidedly French, and its obsession once again is France's frontier. Forty-nine out of eighty-four sheets in the atlas are concerned with Europe, and of these fully eighteen are maps of territory along the eastern arc of the French frontier. This map of Lorraine forms part of that series of frontier maps.

References: Buisseret (1964), King, Konvitz and Pastoureau (1984).

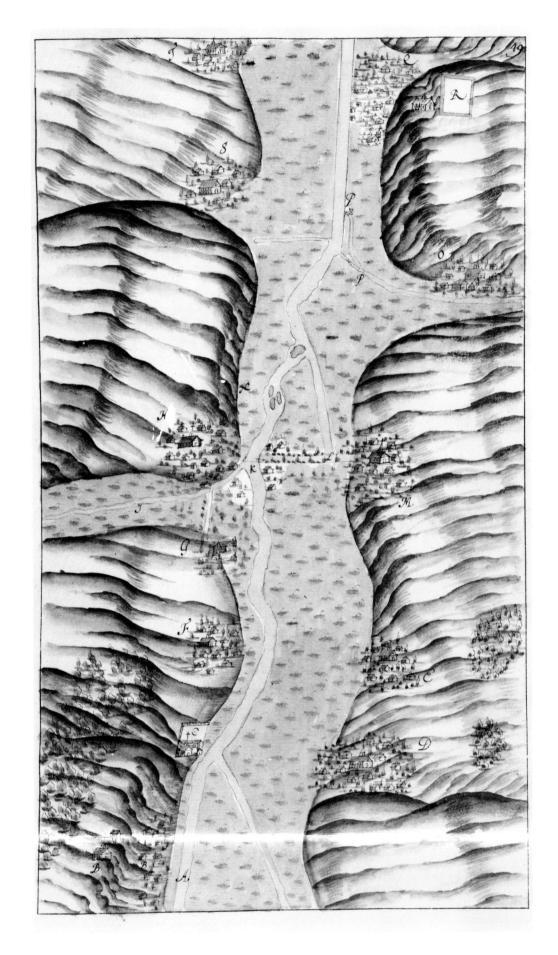

27 Le Sieur Le Nin, Sheet 49 from the Atlas of the Somme (1644) [General Collections, purchased from the Arthur Holzheimer Map Fund and the Samuel R. & Marie-Louise Rosenthal Fund]

28 F. J. Rothhaas, Section from the "Track Book" [of the New Orleans, Jackson, and Great Northern Railroad] (New Orleans, ca. 1859) [Illinois Central Railroad Archives]

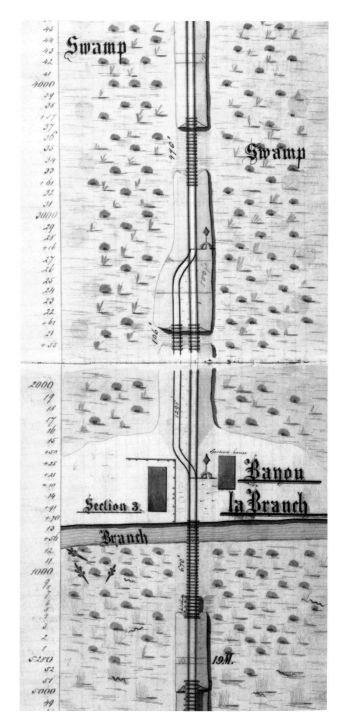

Here are two examples of an ancient cartographic genre, the route map. The atlas on the left was drawn by the sieur Le Nin, a French royal engineer, in 1644. It constitutes a report on the defensive possibilities of the river against attack from the Spanish Netherlands to the north. Le Nin provides detailed maps of the major bridges and fords; this sheet shows a portion of the river just west and north of Amiens, between Belloy and Condé. The Somme was navigable only about as far Amiens, and in Le Nin's time some parts (as at the top of this sheet) had already been straightened. A canal was begun in 1725 and worked on intermittently for over 100 years before final completion. On Le Nin's plan we can see, among other features, the ruins of a Roman camp at "R," a millrace and mill at "P," a ford across the river at "K," and an abbey at "C."

The atlas on the right was the work of an engineering draftsman named F. J. Rothhaas. It represents the main line of the New Orleans, Jackson, and Great Northern Railroad between New Orleans, Louisiana and Canton, Mississippi. The 206 miles of track were laid between 1853 and 1859; Rhothhaas was paid $125 per month in 1857. Using the generous scale of 1:4,800 (one inch = 400 feet), Rothhaas was able to show the right-of-way in great detail and even, like some early explorers, depict local flora and fauna (note the alligators and snakes just below Bayou la Branch).

Although separated by more than two hundred years, by thousands of miles, and by an industrial revolution, it is possible to note characteristics of these two maps which are typical of the genre of route map. Significantly, both are manuscripts and both are drawn by engineers. Most route maps are made for the limited, practical ends of constructing or improving routes, and call for the services of trained engineers rather than dilettante cartographers or commercial map publishers. In both atlases, orientation is irrelevant; the presentation is entirely linear and the individual route segments are simply oriented in the direction of the page. In both atlases, the route is all; detail on either side of the route is minimal and "fades out" with distance from the route. Finally, the persistence of the symbols for trees and marshland or prairie is striking.

References: Blanchard and Buisseret (1990).

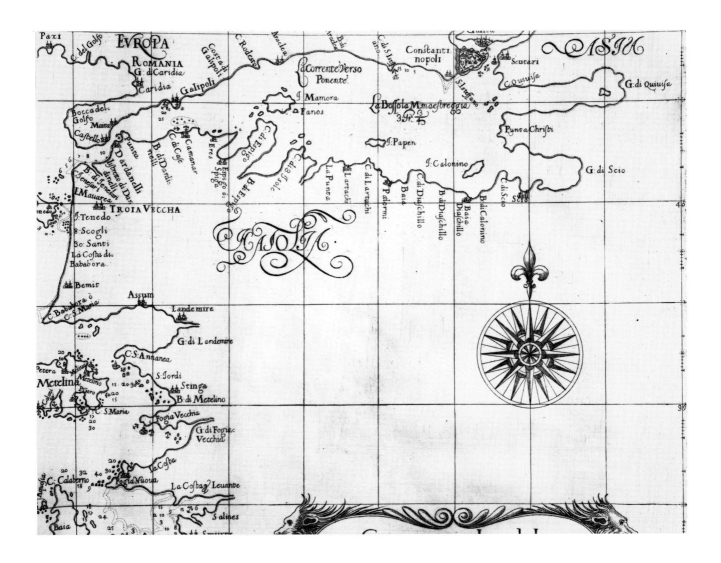

29 Sir Robert Dudley, Detail from the "Carta Particolare del Arcipelago," in the *Arcano del Mare* (Florence 1646-47) [Edward E. Ayer Collection]

30 Gasparo Tentivo, "Canale di Constantinopoli," from a Portolan Atlas (Venice, 1661) [Franco Novacco Collection]

The Tentivo portolan atlas, though written after Dudley's *Arcano*, looks back to an earlier tradition. It retains the windroses of the medieval portolan charts, together with their exaggerated embayments of the coasts, and their elegant bird's-eye views of cities like Constantinople. Virtually the only innovative feature is the inscription of place-names, no longer at right-angles to the coast. Gasparo

Tentivo is a shadowy figure, said to have compiled a chart of the Gulf of Venice and to have collaborated with Vicenzo Coronelli. It is not easy to see for whom his atlas was intended; certainly any seaman who tried to use it in the Dardanelles or its approaches would rapidly have come to grief. Perhaps the atlas was designed to give armchair travellers a feel for the eastern Mediterranean Sea.

At first sight, the corresponding section in the *Arcano* seems a marked improvement. There is no system of windroses, the map is drawn on the Mercator projection, and it has been well printed with a most elegant typography. However, comparison with modern maps of the same area forces the conclusion that Dudley's sources were scarcely better than those of Tentivo; anybody trying to use this chart for actual navigation would soon have perceived that it had been drawn in some study, far from the rocks and capes that it

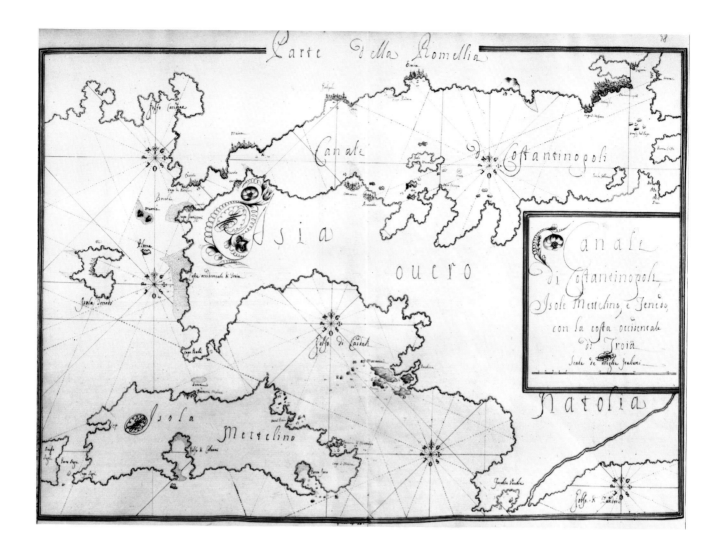

purports to show.

Perhaps we should look upon these two atlases as demonstrations that something better was needed, if maritime navigators were to come to a serious reliance on charts. In Holland there was already a better example (see number 17 of this catalog), and in the next century many countries would begin to produce versions of what the English called "Admiralty charts."

References: Donazzalo and Zavatti.

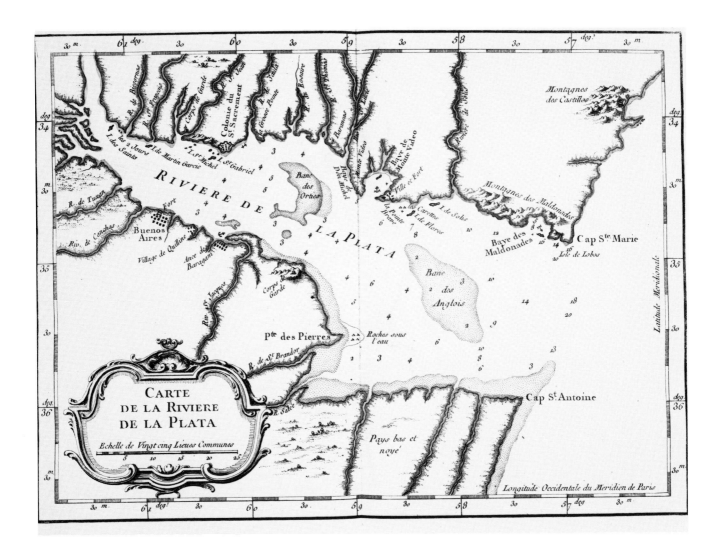

31 Jacques-Nicolas Bellin, "Carte de la Riviere de la Plata," from *Le Petit Atlas Maritime* (5 vols., Paris, 1764) [Edward E. Ayer Collection]

32 Vincent Tofiño de San Miguel, Chart of the Estuary of the Río de la Plata from the *Cartas Esféricas* (Madrid, 1786-1813) [Edward E. Ayer Collection]

As we have seen (numbers 17 and 18), printed atlases of marine charts began to appear in Europe in the later sixteenth century. They eventually included charts of virtually every part of the world that could be reached by water. In the development of these atlases, the French government was pre-eminent, bringing out the first version of *Le Neptune François* as early as 1693. During the eighteenth century, this work was notably carried on in France by Jacques-Nicolas Bellin, one of whose maps we show here. It comes from *Le Petit Atlas Maritime*, a five-volume atlas covering the whole world in about 500 maps. Bellin's sources were varied, and his work often approximate, but its sheer comprehensiveness and extent compel admiration.

The English long continued to rely upon private chartmakers, but some other European powers followed the French in producing official charts. Foremost among these were the Spaniards, who in other respects as well tended to follow the French lead during the eighteenth century. Our second map is from the Spanish Dirección Hidrográfica de Madrid, whose leading cartographer was Don Vincent Tofiño de San Miguel (1732-1795). His *Atlas Marítimo* was the equivalent of the French *Neptune François* or the British *Atlantic Neptune*.

This map shows the same area as the map by

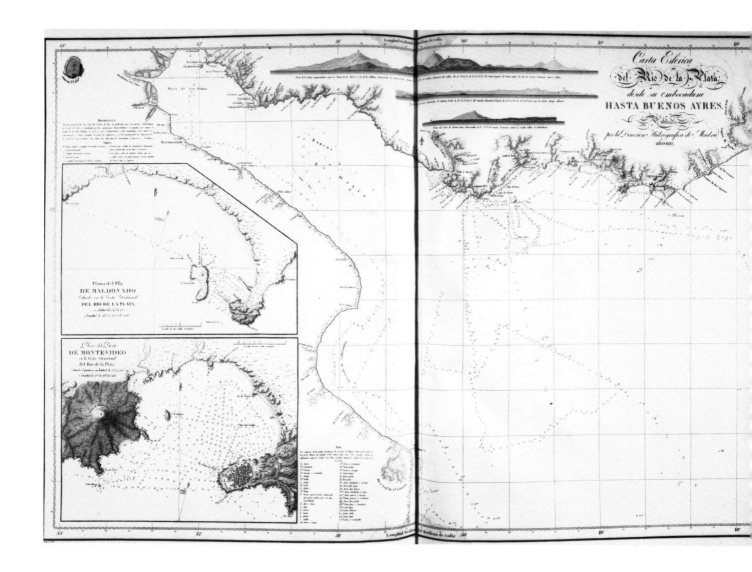

Bellin, but it is considerably more accurate and detailed; indeed, as we examine it in detail, we see that Bellin's delineation of many features was rather gross. Still, the main point about these European atlases is not that they were all superlatively accurate, but that they were so numerous - each vessel in a squadron could carry one - and covered so much of the world. With instruments like these, the European navies and merchant fleets could and did project their countries' power all over the world, giving rise to a period of European hegemony that seems to have ended only yesterday.

References: Garant, Howse and Navarrete.

8e canton. *Les dix-huit mesures.*

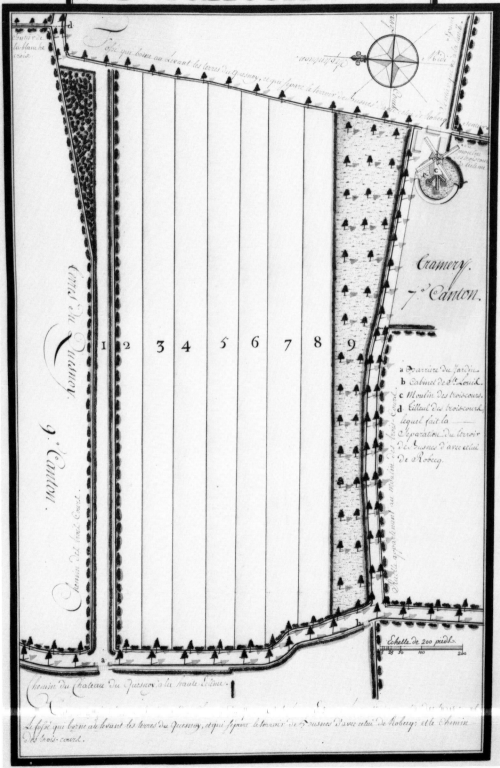

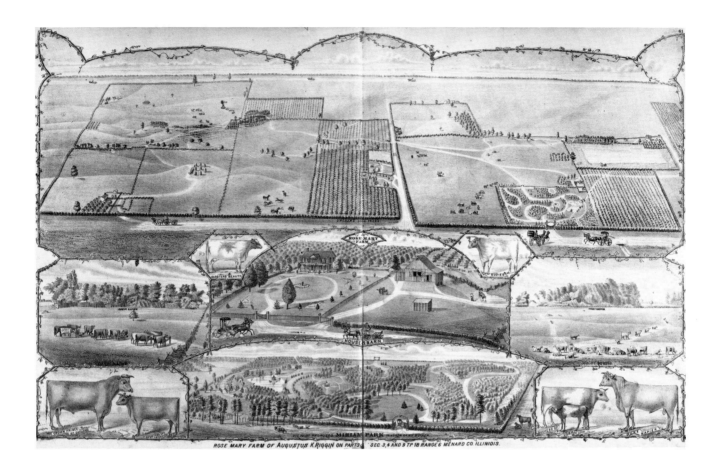

ROSE MARY FARM OF AUGUSTUS K.RIGGIN ON PARTS SEC. 3,4 AND 9 TP 18 RANGE 6 MENARD CO. ILLINIOIS.

33 Christophe Verlet, Plan of the 8th *canton* from the [atlas] "Carte Géométrique et Figurative de la Paroisse de Busnes," (1782) [General Collections]

34 "Rose Mary Farm of Augustus K. Riggin," from W.R. Brink & Co., *Illustrated Atlas Map of Menard County, Illinois* ([Edwardsville], 1874) [General Collections]

These two images of the countryside are separated by less than one hundred years in time, but by a vast difference in the nature of the societies that produced them. Verlet was one of the *arpenteurs*, or surveyors, who seem to have flourished around Lille in the later eighteenth century. His manuscript atlas contains 100 maps, each one showing a "canton," or administrative sub-division. The maps are the same size, but at widely differing scales, showing a great mastery of the surveyor's art.

Verlet's countryside is that of pre-Revolutionary France. There is a superb page devoted to the château of the local *seigneur*, and the buildings of the church are also prominent. On this plan of the eighth canton we see a set of long fields, that probably survive from the strips of medieval times. At the top right is a windmill, the "Moulin des Trois-Cours," as we learn from the letter "c" on the keyed table at the bottom.

The view from the Menard County atlas showing Augustus Riggin's farm also has a mill, but there the parallels end. This is an artefact of the industrial age, lithographed for the mass market. Augustus Riggin no doubt paid well to have Rose Mary Farm so prominently shown, in a work that would have been widely diffused in the county. Although flat, the countryside is marvelously varied, with groves, orchards, pastures and even provision for a park. There is no lord's house, and there are no cottages for peasants either; this is an idealized vision of what Everyman Farmer could aspire to, in the burgeoning lands of the American Midwest.

References: Buisseret (1988) and Conzen.

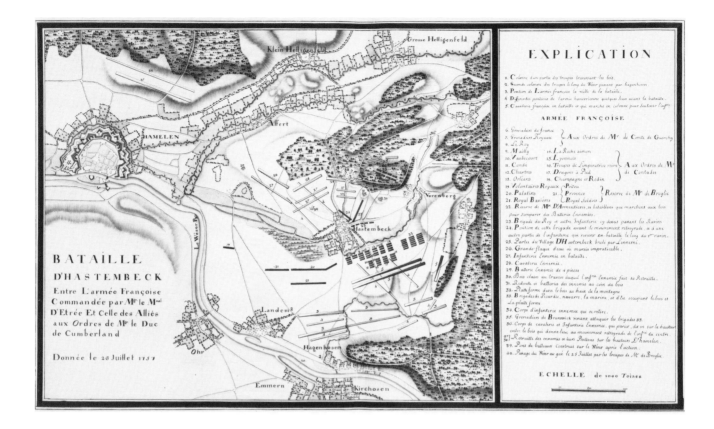

35 Anon., "Bataille d'Hastembeck" (1757) [General Collections]

36 W.C. Wilkinson/William Faden, *Plan of the Encampment and Position of the Army under His Excellency Lieutenant General Burgoyne at Sword's House on Hudson's River near Stillwater* (London, 1780) [General Collections]

The Library possesses two collections comprising over two hundred manuscript and printed battle plans from the eighteenth and early nineteenth centuries, from which these two items are taken. Though one plan is French and one British, one manuscript and one printed, their rendering of relief, forests, fields, and units of the engaged armies is remarkably similar. This is no accident; war among eighteenth-century European powers was a highly regulated and codified affair, and its combatants at all levels were schooled in more or less the same regimen. The military engineers who compiled these plans and the draftsmen who prepared the fair copies of them likely read the same manuals (or translations of them), which laid

out rather stringent rules for rendering particular natural and cultural landscape features.

The manuscript plan depicts the battle of Hastenbeck (1757) near Hameln in northern Germany, a major defeat for Prussia and its Hanoverian allies at the hands of the French army early in the Seven Years' War. A careful reading of the explanation at the right side of the map provides a fairly clear picture of the course of the battle, from the emergence of a French column from a forest at the lower right to the final retreat of the allied forces to a height above Hameln at upper left. Though manuscript, the map is hardly a rude sketch. Drafts made in the field were compiled into intermediate mock-ups called *brouillons*, from which fair copies like this one were made, destined for battlefield reports sent to the proper military authorities.

Finished battlefield narrative maps had obvious attractions that commercial publishers such as William Faden frequently parlayed into personal profit. One of Faden's printed plans is our second exhibit, showing the engagement at Freeman's Farm in New York. It took place shortly before the collapse of the 1777 British campaign aimed at dividing the rebellious North American colonies

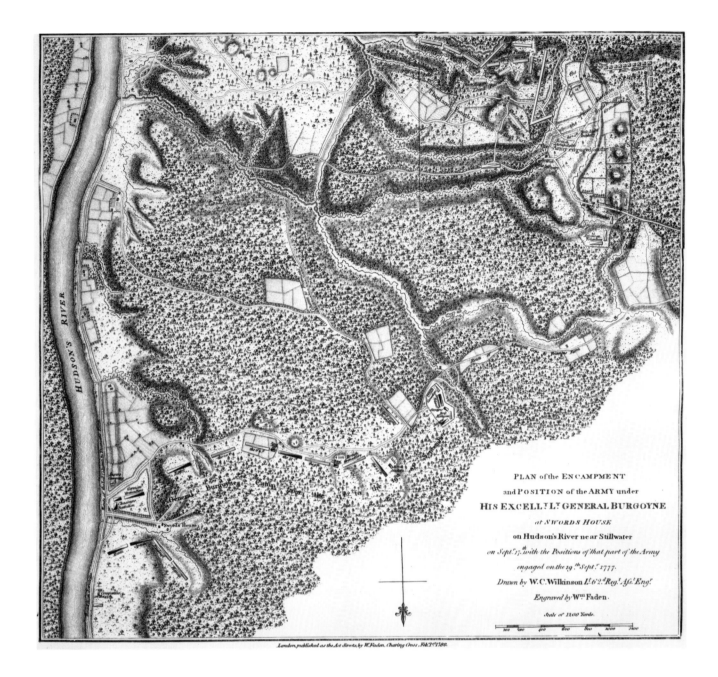

PLAN of the ENCAMPMENT
and POSITION of the ARMY under
HIS EXCELL.Y L.T GENERAL BURGOYNE
at SWORDS HOUSE
on Hudson's River near Stillwater
on Sept.r 17.th with the Positions of that part of the Army
engaged on the 19.th Sept.r 1777.
Drawn by W.C.Wilkinson L.t 6.2.d Reg.t Ass.t Eng.r
Engraved by W.m Faden.

Scale of 1200 Yards.

London, published as the Act directs, by W. Faden, Charing Cross, Feb.y 2.d 1780.

along the Hudson River-Lake Champlain axis. Printed three years after the event, this was not a newsmap, but probably interested armchair military enthusiasts who looked to these maps to supply more information about campaigns and battles with which they were already familiar. In this case the story of the battle is narrated directly on the map, aided by the use of a printed overlay showing the changing positions of the engaged units. The map was also published in Burgoyne's *State of the Expedition from Canada* (London, 1780).

References: Harley and Nebenzahl.

37 Wilhelm Haas, *Nouvelle Carte de l'Italie* **(Basel, n.d.) [General Collections, purchased from the Andrew McNally Fund]**

38 *Rand, McNally and Company's New Railway Guide Map of the United States and Canada* **(Chicago, 1873?) [Rand McNally Collection]**

Literacy rates rose rapidly in the West during the nineteenth century, and with them rose cartographic literacy. The emergence of a new mass market for reference, travel, and newsmaps was a stimulus for experimentation with cheap alternatives to map printing with engraved copper plates which, though versatile and expressive, was both labor-intensive and slow.

One early alternative tried by European printers involved the composition of images entirely from movable type. We show a map of Italy by Wilhelm Haas (the younger), of Basel, who specialized in this technique. Every road, river, town, mountain, and coastline on this map has been printed from metal type composed in forms like those used to print the pages of books. One advantage of this technique was that it allowed letters and graphic symbols to be printed simultaneously. The pieces of type, however, could not be made easily to touch one another, and linear marks, for example, frequently looked both stilted and non-continuous. In short, the difficulties of developing an exhaustive set of interchangeable characters for complex landscapes proved insurmountable, and typographic cartography was abandoned by the middle of the nineteenth century.

Wax engraving (or cerography) was ultimately much more successful in meeting the needs of large commercial cartographers, especially in this country during the last few decades of the nineteenth century and the first few decades of the twentieth century. Despite the name, this was a relief printing process, involving the engraving or punching of a wax mold, from which was made a copper-clad metal plate with the image in relief. The printing plates were extremely durable, yet

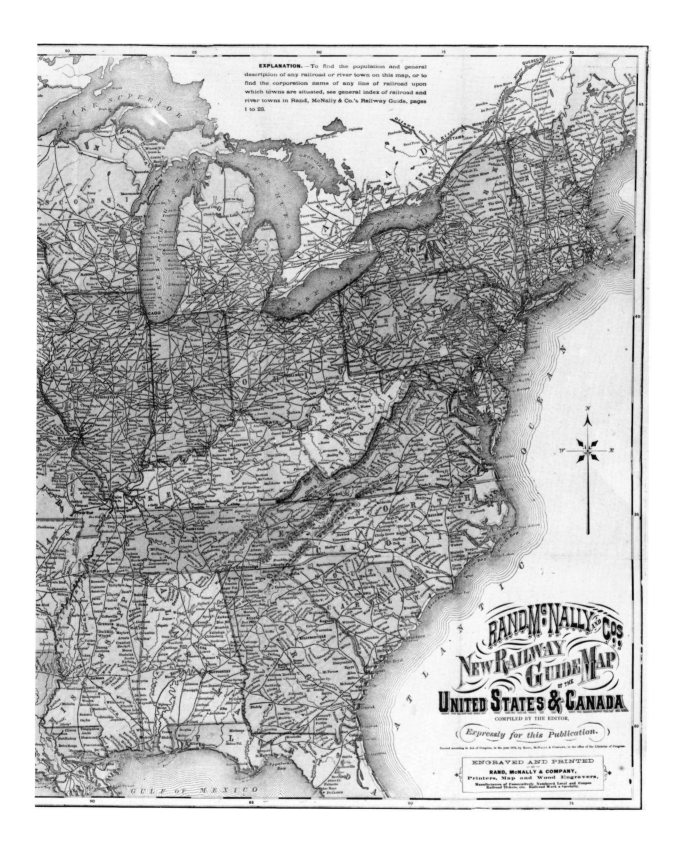

could be finely detailed. Printing in multiple colors by this method was relatively simple and inexpensive, and thus commercial printers like Rand McNally & Co. found it useful for the manufacture of their many cheap but appealing atlases and travel maps. We show one of Rand McNally's earliest maps, a railroad map of the United States published for the firm's railway travellers' and shippers' guide in 1874.

References: Hoffmann-Feer, McMurtrie, Peters and Woodward (1975) and (1977).

39 Ordnance Townland Survey of Ireland, detail from Sheet 37, County Down (Dublin, 1834) [General Collections]

40 "Provincia Ecclesiastica di S. Luigi," from *L'Orbe Cattolico ossia Atlante Geografico-Ecclesiastico* **(Rome, 1858-1859) [General Collections, purchased from the Andrew McNally Fund]**

It was during the nineteenth century that large-scale printed surveys became common in the countries of Europe and in their overseas possessions. In some ways, Ireland almost qualified as a British overseas possession; as Lord Salisbury

put it in 1863, "the most disagreeable part of the three kingdoms is Ireland, and therefore Ireland has a splendid map." This map had been compiled between 1824 and 1846, and at a scale of six inches to the mile (1:10,560) needed more than 1500 sheets.

As our example shows, they were superbly engraved on copper (at the very end of the copper-engraving era) and could show a remarkable amount of detail. In Downpatrick, for instance, at the upper right, we can detect the "English," "Irish," and "Scotch" Streets, each with its appropriate place of worship. A huge gaol seems to brood over the town, and in the countryside numerous antiquities are carefully marked. This was a tremendous map for raising taxes or marching troops - or for taking students on field-trips. Its very large scale was unprecedented, when England was mapped at

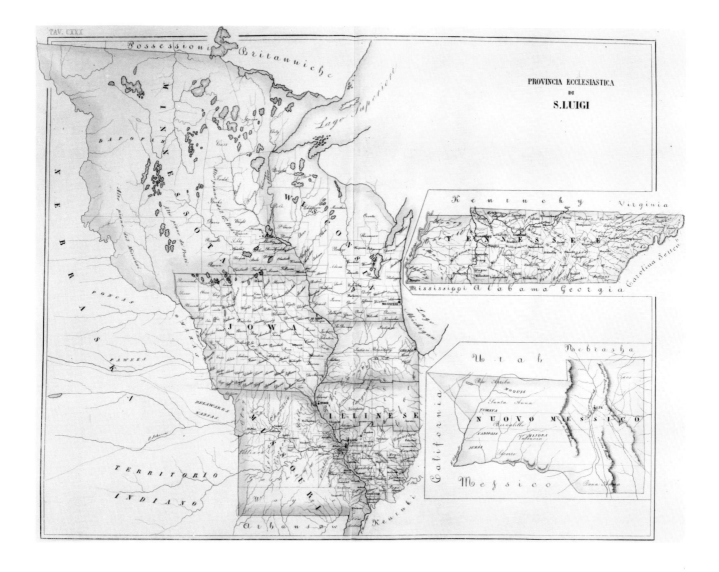

1:63,360, France at 1:80,000, Sweden at 1:100,000 and so forth.

The atlas volume from *L'Orbe Cattolico* - which seems to be very rare - testifies to a different kind of desire to pull things into order cartographically. It contained 138 plates, produced by the new lithographic process, depicting the church's provinces as seen from Rome. In the past, individual orders like the Jesuits had produced maps of various regions; now for the first time the Pope could visualize all his archdioceses and dioceses throughout the world in one volume with a uniform format.

The atlas is open at the Saint Louis province, the "Provincia ecclesiastica de S. Luigi," and shows the states of the newly-defined Midwest abutting the "Territorio Indiano." This map is in exactly the same style as the others in the atlas, with the dioceses marked with a red spot, and the "provinces" shown in pastel shades. The general effect of the atlas is thus to lend a sense of order and unity to the Pope's actually very disparate world, just as the Ordnance Townland Survey imposes an elegant uniformity over Ireland's contention-ridden countryside.

References: Andrews.

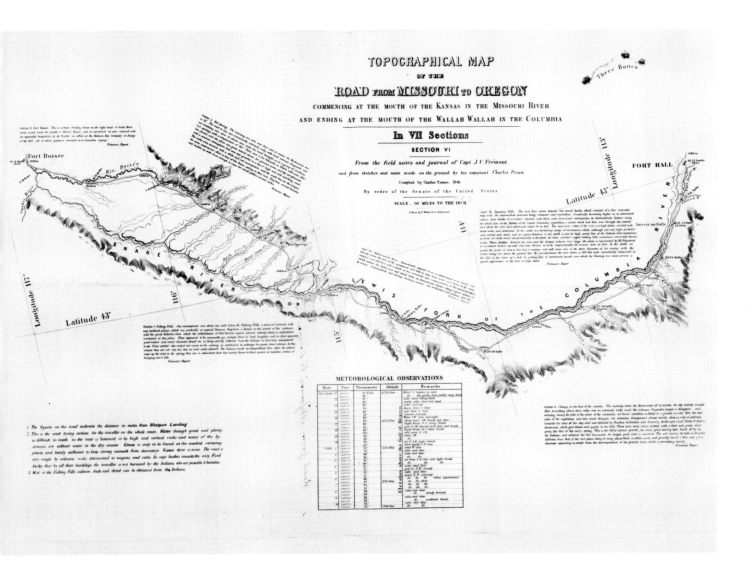

41 Charles Preuss, *Topographical Map of the Road from Missouri to Oregon* **(Baltimore, 1846) [Everett D. Graff Collection]**

42 George T. Robinson, Detail from the "Map of the Military District, Kansas and the Territories" (n.p., 1866) [General Collections, Gift of Arthur Holzheimer]

These two maps are linked by their depiction of the Oregon Trail and the Transcontinental Railroad, two legendary pathways taken during the conquest of the West. The overland route from Independence, Missouri (near Kansas City) to the Oregon Territory via South Pass in western Wyoming had been known to intending settlers for more than two

decades, but no detailed map of the route (actually a braided system of trails) existed before the military engineer Charles Preuss committed this one to paper at the request of the U.S. Senate. The map shows the path of John Charles Fremont on his second expedition, which included Preuss, in 1843. Acclaimed for its utility to potential travellers along the trail, its artful rendering of the trailside landscape, together with excerpts from the expedition's log, give us some idea of the difficulty of the journey.

The Oregon Trail is but one of the many trails and expedition routes crisscrossing the Western landscape of George Robinson's map of the central Great Plains and Rocky Mountains made for General Grenville Mellon Dodge. Compiled just after the upheaval of the Civil War had ended, this map shows the national energy once again

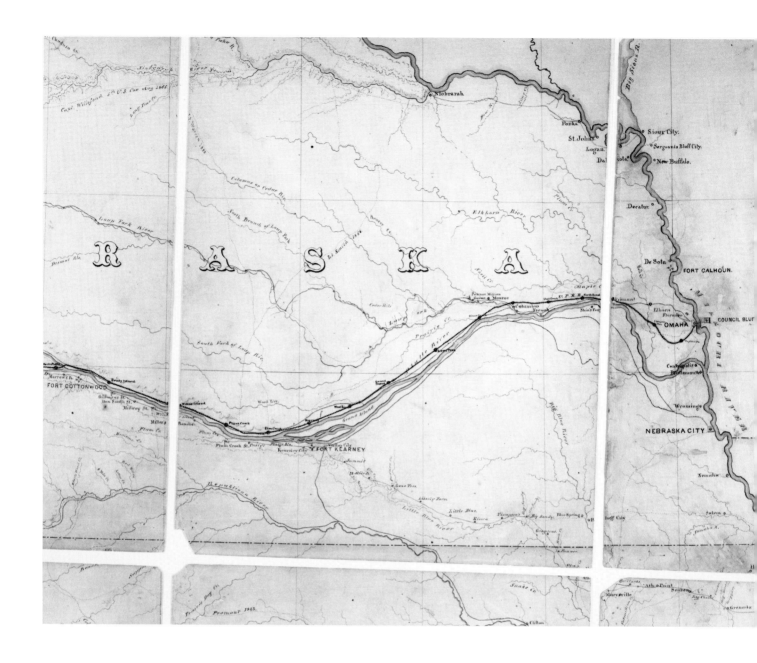

focussing on its western territories. Drawn in the style of the military engineers of that time, the map was made for Dodge when he was commander of the military district of that part of the country. It followed him when he left the army to take the post of chief of engineers for the Union Pacific Railroad.

The first stretch of this famous road is sketched in from Omaha as far as North Platte, Nebraska, and contains pencilled additions and erasures that seem related to further construction.

References: Thurman and Wheat.

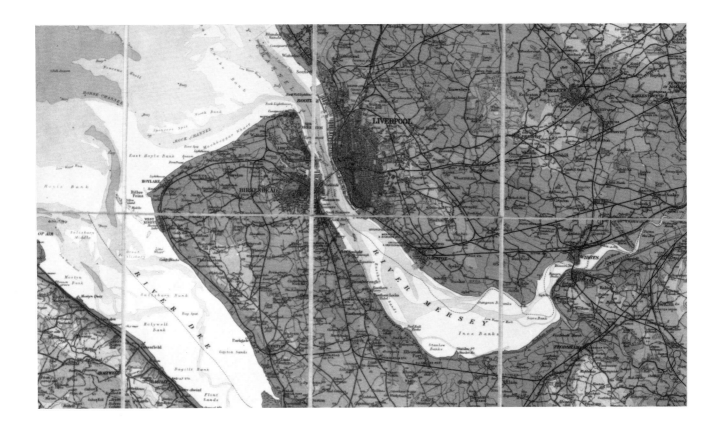

43 Detail from *Bartholomew's Half-Inch to Mile Map of England and Wales*, sheet 8, Liverpool and Manchester (London, c. 1907) [General Collections, purchased from the Andrew McNally Fund]

44 Rand, McNally *Official 1923 Auto Trails Map, district no. 13* (Chicago, 1923) [Rand McNally Collection]

These road maps belong to the early era of automobile travel in Europe and North America, when there were few paved roads and more beasts than engines travelling over them. The richly colored Bartholomew "half-inch" map is one of a set of 37 covering all of England and Wales like the topographic maps of the Ordnance Survey, upon which it is based, the map served many kinds of travellers. The delineation of the region's dense network of interurban railways, for example, is a reminder that trains were still the superior mode of overland travel in almost every way. The map's considerable attention to topography and road surface quality is also understandable given the low horsepower and fragility of bicycles and early motor cars. Versatile maps like this one remained the preferred type of road map for many decades after the arrival of the automobile, in testimony to Britain's leisurely and relatively restricted domestic travelling habits.

In contrast, Rand McNally's 1923 "Auto Trails" map of the northern Rocky Mountains of the U.S., a forerunner of the free oil company road map, was designed particularly for automobile use. Printed in only two colors, it has nevertheless a spare elegance suited to its single-minded utility. Topographical representation is minimal, and makes the Rockies seem mere hills, a not entirely accidental effect, since states, automobile and oil companies, road builders and map makers all emphasized the ease and independence of automobile travel. The region's important network of railroads is missing altogether - though we may trace the route of the Union Pacific by the road paralleling it, routes "10" and "34" across southern Wyoming. Even at this early date, the American affection for rapid, long-distance automobile pleasure trips to scenic wonders is written all over the map.

References: Akerman, Nicholson and Ristow.

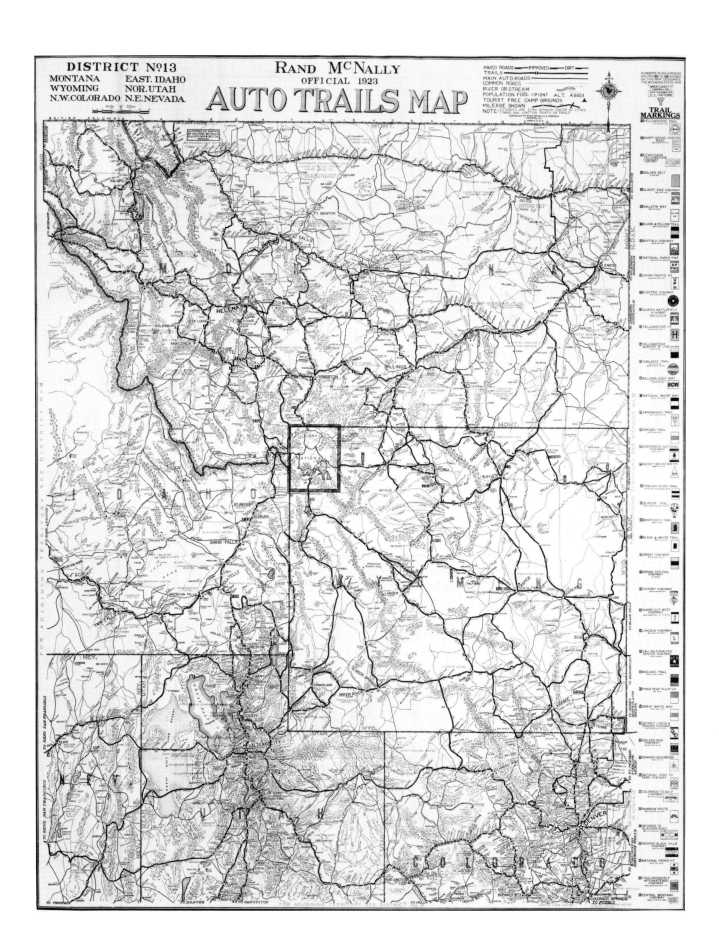

Bibliography

Akerman, James, "Selling Maps, Selling Highways: Rand McNally's 'Blazed Trails' Program," *Imago Mundi*, 45 (1993)

Almagiá, Roberto, *Monumenta Italiae Cartographica* (Florence, 1929)

Andrews, John H., *A Paper Landscape: The Ordnance Survey in Nineteenth-Century Ireland* (Oxford, 1975)

Blanchard, Anne, *Les Ingénieurs du "Roy" de Louis XIV à Louis XVI* (Montpellier, 1979)

Bosse, David, "Johann-Gabriel Sack and his Maps," *Mapline*, 26 (June 1982) 1-4

Breydenbach, Bernhard von, *Die Reise ins Heilige Land: Ein Reisebericht aus dem Jahre 1483*, ed. Elisabeth Geck (Wiesbaden, 1977)

Brown, Lloyd A., *The World Encompassed* (Baltimore, 1952)

Buisseret, David, "Les Ingénieurs du Roi au Temps de Henri IV," *Bulletin de la Section de Géographie*, lxxvii (1964) 13-84

Buisseret, David, *Rural Images: The Estate Plan in the Old and New Worlds* [exhibit-catalog] (Chicago, 1988)

Buisseret, David, "Newberry Acquisitions," *Mapline*, 57 (March 1990) 6-7

Buisseret, David, *Mapping the French Empire in North America* [exhibit catalog] (Chicago, 1991)

Campbell, Tony, "Census of Pre-Sixteenth-Century Portolan Charts," *Imago Mundi*, 38 (1986) 67-94

Campbell, Tony, *The Earliest Printed Maps* (Berkeley, 1987)

Caraci, Giuseppe, "The Italian Cartographers of the Benincasa and Freducci Families and the So-Called Borgiana Map of the Vatican Library," *Imago Mundi*, 10 (1953) 23-49

Cohen, Simon, tr., *Historical Essay on the Colony of Surinam*, 1788 (Cincinnati, 1974)

Cortesão, Armando and Avelino Teixeira da Mota, *Portugaliae Monumenta Cartographica* (6 vols., Lisbon 1960; second edition 1987)

Conzen, Michael P., "The County Landownership Map in America: Its Commercial Development and Social Transformation, 1814-1939," *Imago Mundi*, 36 (1984) 9-31

Dictionnaire Géographique et Administratif de la France (Paris, 1905)

Dizionari Biografico degli Italiani, 33 (1987) 35-44, articles on "Gregorio Dati" and "Leonardo Datti"

Donazzolo, Pietro, "Di un Portolano Inedito e Sconosciuto Reguardante il Mediterraneo e Specialmente l'Egeo di 'Gasparo Tentivo,'" *Real Società Geografica Italiana, Bollettino*, ser. 6, v. 3 (1926) 847-853

Errera, C., "Carte e Atlanti di Conte di Ottomano Freducci," *Rivista Geogr. Italiana*, 2 (1895) 237-241

Ganong, William F., "The Identity of the Animals and Plants Mentioned by the Early Voyagers to Eastern Canada and Newfoundland," *Transactions of the Royal Society of Canada*, 3rd ser., 3, sec. 2 (1909) 197-202

Garant, Jean Marc, *Jacques-Nicolas Bellin (1703-1772): Cartographe, Hydrographe, Ingénieur du Ministre de la Marine: Sa Vie, Son Oeuvre, Sa Valeur Historique* (Montréal, 1973)

Harley, J. B., et al., *Mapping the American Revolutionary War* (Chicago, 1978)

Harley, J. B. and David Woodward, eds., *The History of Cartography* (vol. i, Cartography in Prehistoric, Ancient and Medieval Europe and the Mediterranean) (Chicago, 1987)

Heidenreich, Conrad E., *Explorations and Mapping of Samuel de Champlain, 1603-1632* [Cartographica Monograph, 17] (Toronto, 1976)

Hoffmann-Feer, *Die Typographie im Dienste der Landkarte* (Bascl, 1969)

Howse, H. D. and M. Sanderson, *The Sea Chart* (Newton Abbot, 1973)

Karrow, Robert W., Jr., *Maps and Mapmakers of the Sixteenth Century* (Chicago, 1993)

King, James E., *Science and Rationalism in the Government of Louis XIV, 1661-1683* (Baltimore, 1949)

Kish, George, *The Suppressed Turkish Map of 1560* (Ann Arbor, 1957)

Koeman, Ir. Cornelis, "The *Theatrum Universae Galliae*, 1631: An Atlas of France by Joannes Janssonius," *Imago Mundi*, 17 (1963) 62-72

Koeman, Ir. Cornelis, "Bibliographical Note" in *Gerard de Jode, Speculum Orbis Terrarum, Antwerpen 1578*, v-xiv. Theatrum Orbis Terrarum Series of Atlases in Facsimile, ser. 2, vol. 2 (Amsterdam, 1965)

Koeman, Ir. Cornelis, *Joan Blaeu and His Grand Atlas* (Amsterdam, 1970)

Koeman, Ir. Cornelis and H. J. A. Homan, eds., *Atlantes Neerlandici* (6 vols., Amsterdam, 1967-1971; Alphen-aan-den-Rijn, 1985)

Konvitz, Josef W., *Cartography in France, 1660-1848: Science, Engineering and Statecraft* (Chicago, 1987)

Kraus, Hans P., *Sir Francis Drake: A Pictorial Biography* (Amsterdam, 1970)

McMurtrie, Douglas C., *Printing Geographic Maps with Movable Types* (New York, 1925)

Mollat du Jourdain, Michel and Monique de La Roncière, *Sea Charts of the Early Explorers* (New York, 1984)

Navarrete, Martin Fernandez de, "Vicente Tofiño de San Miguel," in *Biblioteca Maritima Española* (Madrid, 1831)

Nebenzahl, Kenneth, *Bibliography of Printed Battle Plans of the American Revolution, 1775-1795* (Chicago, 1975)

Nicholson, T. R., *Wheels on the Road: Road Maps of Britain 1870-1940* (Norwich, 1983)

Nordenskiöld, A. E., *Periplus: An Essay on the Early History of Charts and Sailing-Directions* (Stockholm, 1897)

Pastoureau, Mireille, *Les Atlas Français XVIe-XVIIe Siècles: Répertoire Bibliographique et Etude* (Paris, 1984)

Pastoureau, Mireille, ed., Nicolas Sanson d'Abbeville, *Atlas du Monde, 1665* (Paris, 1988)

Peters, Cynthia, "Rand McNally and Company in the Nineteenth Century: Reaching for a National Market," in Michael P. Conzen, ed., *Chicago Mapmakers: Essays on the Rise of the City's Map Trade* (Chicago, 1984)

Pickering, Charles, *Chronological History of Plants* (Boston, 1879)

Ristow, Walter W., "A Half Century of Oil-Company Road Maps," *Surveying and Mapping*, 24 (1946) 617-637

Schulz, Juergen, *The Printed Plans and Panoramic Views of Venice (1486-1797)* (Florence, 1970)

Schulz, Juergen, "Jacopo de Barbari's View of Venice: Map Making, City Views, and Moralized Geography before the Year 1500," *Art Bulletin*, 15 (1978) 425-474

Shirley, Roderick, *The Mapping of the World: Early Printed World Maps, 1472-1700* (London, 1983)

Skelton, R.A., "Bibliographical Note," in Gerard de Jode, *Speculum Orbis Terrarum, 1587* (Amsterdam, 1965)

Skelton, R.A., "Bibliographical Note," in Francesco Berlinghieri, *Geographia, Florence, 1482* (Amsterdam, 1966)

Smith, Clara A., *List of Manuscript Maps in the Edward E. Ayer Collection* (Chicago, 1927)

Thurman, Mel, "Warren, Dodge and Later Nineteenth-Century Army Maps of the West," *Mapline*, 53 (March 1989) 1-4

Waghenaer, Lucas Janszoon, *Spieghel der Zeevaert* (Leyden, 1584-1585), ed. R.A. Skelton (Amsterdam, 1964)

Wheat, Carl I., *Mapping the Trans-Mississippi West 1540-1861* (5 vols., San Francisco, 1957-1963)

Wolff, Hans, ed., *Philipp Apian und die Kartographie der Renaissance* [Bayerische Staatsbibliothek Ausstellungkataloge, 50] (Munich, 1989)

Woodward, David, *The All-American Map* (Chicago, 1977)

Woodward, David, ed., *Five Centuries of Map Printing* (Chicago, 1975)

Zavatti, Silvio and Franco, "L'Arcano del Mare di Robert Dudley," *L'Universo*, 53 (1973) 695-712